AXIOMS for the INVENTOR

Twenty Two Tips to Produce Patentable Inventions

Charles C. Rayburn

ISBN: Softcover   978-1-4990-5526-9
        EBook   978-1-4990-5527-6

Rev. date: 07/30/2014

To order additional copies of this book, contact:
Xlibris LLC
1-888-795-4274
www.Xlibris.com
Orders@Xlibris.com

PREFACE

What is an invention? Are all inventions patented? Is an inventor someone who has patents issued? How are potential inventors identified?

Industrial managers are constantly wondering how their company can build a better product and upstage their competition. They seek inventions. They wonder if there are fledgling inventors in their midst just waiting to receive that flash, that lightning bolt which will forever alter the future of mankind.

During the past 35 years, I've worked as an industrial manager and inventor averaging just over two U.S. Patents per year. My failures outnumber my successes. Fortunately, my successes have far more than subsidized my failures. My bosses have been tops, always supportive, willing to carry me through the lean times, hoping that I will hit a home run.

The purpose of this writing is to assure the reader that we are all inventors. There are no proven scales for measuring the potential for invention. Is there a test for babies to indicate who will accumulate wealth over their lifetime? Probably not. The same is true for inventing. Wealth accumulation and successful inventing are each directed by interest, effort, and opportunity. My interests and efforts were about average, but my opportunities were abundant.

For easy reading I have condensed various points, printing them in bold type. Some commentary accompanies each to explain the inclusion of the bold statement.

Finally, let me assure you that inventing is a game. It is show biz, technology, law, marketing, manufacturing, internationalism, and science rolled into one. If one has a streak of luck, it can be financially rewarding and personally satisfying.

TABLE OF CONTENTS

CHAPTER 1—TO BECOME AN INVENTOR

1:1 Know natural scientific laws, Understand commercial practices and materials.

Today's engineering degrees in general, provide an adequate background in mathematics, physics, chemistry and other sciences to equip the inventor. Commercial practices and materials are not taught in the concentrated college atmosphere and are usually learned only in areas of interest. For example, one operating a plastic extruder may know little about clear white pine, while the sawmill operator knows less about biaxially oriented polymer, but each is well versed in his own specialty.

But, sometimes this happens:

An associate folded a piece of plastic and with scissors, starting at the folded edge, cut out a semi-circular arc. He opened the fold, revealing a round hole in the plastic sheet. He forced a beer can into the under-sized hole. The can was retained by the plastic. Then he cut five additional holes.

This six-pack carrier is now made world-wide by the billions. A billion dollar idea, which required little academic knowledge to conceive and demonstrate. Nonetheless, this inventor has good scientific credentials and stacks of money-making inventions.

Frequently, someone with minor scientific understanding proposes an idea which requires repeal of the 2nd law of thermodynamics, or a momentary dismissal of Heisenberg's Uncertainty Principle. Whether a new idea is simple or complex, it is reassuring if its originator has scientific integrity.

1:2 Find the cutting edge of a targeted technology. Know the issued patent claims. Understand the economics and direction of the technology.

Sixty years ago much of today's technology was buried in obscure laws of physics. The Piezoelectric effect, incompressibility of liquids, and the transmission of sound were well understood. But only in recent times does a backwoods expectant mother come to town in the family pick-up to take an ultrasonic photograph of her in-utero occupant. Technology advances in spurts, then hesitates, then repeats. Unless one works and reads the current literature, making a significant contribution in an existing advanced technology is unlikely. However, principles developed in other fields are often transferable.

As preparation to innovation, one should take time to understand the patents in the target field. Patents expire a fixed number of years after issue and are then available to the public without licensing. Other patents may contain dozens of pages but only one weak claim. Legally, the claim is important. The other pages teach the invention to the public. If your new idea avoids all active claims, you are clear to proceed.

The time allotted each of us for invention is something less than one lifetime. Few researchers work on more than four major projects during their career. More frequently, we are on the cutting edge of a technology only once. Should that technology be a method for producing an outdated article, the chances are low for filing a patent application because it may have been previously treated in patents or there is little economic incentive to pursue it.

Finding the cutting edge of new technology with strong economic implications can be so time consuming that little time remains to contribute to the field. For example, to some extent this is currently true for super conductivity.

1:3 Advocate a novel solution to a problem and fail. Then, the stage is set for real invention.

The psychologists must surely have a field day with this one. For years I was unaware of this phenomenon. Still, I don't understand it.

As inventors we must be willing to plunge headlong into unchartered waters. We risk capsize and the loss of our provisions. It is embarrassing to be seen wading ashore. Our itinerary is scraped. Our promises are broken. Defeat crowds in.

On the bank, dripping wet, we face two choices; either accept the defeat in shame, or regroup and try again with intense determination. Experience tells of circuitous routes. Persistence drives us to find them.

Inventors use *fear of failure and hope of success* more effectively than other professions. Blame cannot be placed on others. All energy must be focused on the objective.

1:4 Don't expect accolades for innovation. New ideas are resisted by those whose job it is to make, sell, and use them.

Ego surely plays a role in motivating the inventor. It's pleasing to see one's invention in every car or telephone. Those who supported the embryonic idea and contributed to its development are greatly appreciated. A moment of personal recognization has all the thrills of an academy award.

Meantime, there is a production problem. "I told you so" is whispered. Salesmen return from the field with scars about the head and shoulders. It will never break into commerce. Customers are having problems. You are responsible for stopping the assembly lines of General Motors and AT&T, all in the same day. Closer examination shows the situation to be less serious than originally reported.

Finally, the new product is moving into the market without technical problems. Demand is exceeding supply. But, the inventor sleeps restlessly, dreaming that one day all cars stopped and all phone circuits opened because the invention quietly carried a fatal flaw.

CHAPTER 2—FINDING AREAS OF INTEREST

2:1 Apply new technology to solve old problems. This can initiate new industries.

Laser technology is a prime example. Imagine, a tiny photon, a minute bundle of light, being powerful enough to cut through thick plates of steel! Also, delicate enough to perform microscopic surgery deep within the eye. Or, capable of such discrimination that the beam can evaporate aluminum from a supporting plastic film without melting the film. Aluminum evaporates at about 2500°C and the supporting plastic melts at 256°C.

The inexperienced inventor need not be concerned about mastering laser physics, or, that someone else found the laser to be very effective for cutting fabric and replacing the surgical scalpel. Instead, one may direct attention to an old process or product which can be improved through using the laser properties. Many industrial processes have remained relatively unchanged since 1930 or earlier. High capital investments, in such cases, protect the status quo. New technology may completely alter the method and article and offer patent protection to reduce the new investment risk.

2:2 Greatest reward is derived from solving a problem which was thought to be insolvable especially if the solution fills a large market need.

Many hours have been devoted to somehow breaking down water into its hydrogen and oxygen content with less energy than liberated when they reform. A large market is ready and waiting. Free energy, a clean environment, new modes of travel, new building climate control, the advantages are sufficient to warrant the denial of entropy.

The thought of developing film immediately after the picture was taken was a radical idea. Everyone knew a dark room and pans of solutions were absolutely necessary. No one would waste time on an idea so ridiculous as packaging a dark room into a camera!

On any given day I don't mind making the bed cleaning the floor, dusting the furniture, washing the dishes or even ironing a shirt. Must our grandchildren also spend their lives on such trifles? Some properly take great pride in making a bed or seeing their reflections in their squeaky clean dishes. They will be difficult to convert to a dustless house or a collarless shirt which requires no ironing. Their great grandchildren will pay high prices to return to the bed canopy.

Tradition, with all of its merit, can stifle invention. But a good idea, properly promoted can overcome tradition. Frequently, these cases provide a large waiting market.

2:3 Select a high usage product which has commodity status, then, provide a unique substitute of lower cost and higher quality.

A commodity was once defined as a product of agriculture or mining. Now it has been extended to include high quantity, low cost items of commerce, generally participating in an international economy. Passive electronic components are good examples of this category.

Specifically, the electrostatic film type capacitor was a perfect case for applying this technique of innovation. Most film dielectric capacitors are wound, a manufacturing method which dates back to the beginning of radio. These capacitors are used by the billions in computers, automotive controls, telecommunications, instruments, and all other electronic markets.

A new system was developed which allows dielectrics to be referenced in stick form rather than in the wound convolute form. This produced a new commodity of lower cost, higher quality and smaller size.

The improved economics coupled with smaller size allows this new capacitor to also take portions of ceramic and tantalum capacitor markets which were not available to the large wound capacitor.

For reducing investment risk in developing a new commodity type product, the economic improvement should ultimately cut the commodity's cost in half while providing additional incentives for conversion. One must never underestimate the power of existing investment for producing commodities by present methods. Accountants can play tricks to discourage investment in new commodity manufacturing equipment in contrast to continuing with present methods. Generally, the accountants are quite correct.

Intrinsic material cost is the most important consideration when viewing a commodity replacement idea. Present trends lead to the conclusion that escaping to low labor areas is a poor foundation for building a long term business. Mechanization and automation are better long term answers to labor reduction. But material cost follows and often increases in areas of low labor cost. In evaluating intrinsic material cost, study such diverse areas as supply, geography, political control, world abundance, and the energy required to bring the material from its natural state through its conversion and into the consumer's hands.

CHAPTER 3—GETTING TO THE POINT

3:1 Don't waste time working on inventions which are unprotectable through either patents or trade secrets.

Superficially, the above statement appears trivial. Yet, this distinguishes the successful inventor from the tinkerer, the winner from the loser.

Almost everyone knows one who has a latent invention tucked away. It's either a 6-UP story which needed one more try, or it's too late because it's now on the market making another person fabulously wealthy. Closer observation usually reveals that there was never any element of the idea which was protectable.

Conversely, once one has a valid reason to believe that an idea is protectable, the invention may be reaching a mature state. To be certain of protectability before getting started is to eliminate dreams, an unwise option.

Years ago, *machines* were the primary inventive objectives. The sewing machine, cotton gin and similar mechanisms had broad social implication. Labor patterns were altered. Efficiencies increased. Strong muscles were no longer needed. For other reasons we now want strong muscles. Some inventors are busily developing machines to help us build strong muscles.

For most inventors, the *apparatus* has given way to the *method* or *article*. The apparatus is frequently more difficult to communicate to the patent examiner than the method or article. It's protection is more difficult to police. Its uniqueness is generally more difficult to defend.

To avoid wasting time on unprotectable ideas, know the active and inactive claims in the area of interest. Know the leading edge of the applicable technology and Science. Know who is working in the area both nationally and internationally. Then, avoid well trodden paths.

3:2 Prepare technically and economically to understand a need; Then, relax with a distraction and listen for the mind to yield a unique solution to fill the need.

The first word of the above statement is the key, *Prepare.* Without months or years of preparation this scheme does not pay off. Failure in prior attempts to fill a need is excellent preparation.

Why do ideas pop up when least expected? Focused concentration appears to drive away the spark of an idea. Faint stars are more observable in the peripheral vision, then disappear when viewed directly.

Contrary to popular perception, I do not find a brook, stream, or bucolic meadow to be a helpful setting for inventing. It is so serene, and beautiful that it demands attention. My favorite places to listen to the mind are a noisy air terminal after I have checked in, a crowded plane in the center seat between stout people, or at home in a comfortable chair watching a boring TV program.

3:3 Move quickly. New technical ideas cannot be canned and preserved. They perish, in or out of the can.

Procrastination does not belong in the inventor's lexicon. In fact, I can't remember any successful inventor who decided to put off an invention. To the contrary, inventors are more likely to make pests of themselves, asking for something special to be completed yesterday.

Once an idea is born it should be rushed to the nursery for observation. A fresh idea grows stale. No one listens as a new idea is discussed for the second time. A perfectly good idea will die in one pass through a committee.

I vividly remember a staff meeting in 1958 in which a new product idea was shown. More immediate problems caused the staff to pay little attention to the new product. It was then shelved and in desperation pulled out four years later. At that time it was seriously considered and later started into production. Ten years afterwards it was leading its field in the market; however, four big years were wasted.

On this theme of circumventing the committee and moving rapidly, insight is gained into why large bureaucratic corporations are deprived of inventions from within. Extending further, the government is even less likely to yield an original idea. Fortunately, some large corporations are taming the bureaucratic control to provide sheltered nesting areas for fledglings.

A further reason to move quickly results from the fact that every leading edge of every technology is moving. In one day an idea may move from futuristic to historical. Even two futuristic ideas competing for the same funding favor the most advanced in time and concept.

CHAPTER 4—GOOD INVENTING PROCEDURES

4:1 Beware of complicated solutions. They will be too expensive and difficult to communicate.

Typically, the evolution of an apparatus may be divided into the following phases:

Conceptual	5%
Confirmation	7%
Design	28%
Construction	40%
Development	20%
	100% Total Cost

Most of the novelty or invention takes place in the conceptual phase. Often the confirmation phase reveals a problem which requires revisiting the conceptual phase. The purpose of the confirmation phase is to conduct experiments in uncertain areas to reduce the high expense risk. Failure to remove risk in the confirmation phase moves the risk to the development phase. An inexperienced experimenter rushes through the first four phases in the hope of saving time and cost. Then, the development phase may well exceed the expense of the first four phases, often leading to abandonment of the original concept.

Simplification of an idea should always occur in the conceptual phase. The conceptual phase of an article starts with the article and proceeds through each method and apparatus which takes it to its point of need, the customer. If any element of the article-method-apparatus chain is not reduced to an optimum simplicity, the system is open to challenge and loss to competition.

This is an old adage that "Nature sides with a fault." That may be extended to say that "Nature sides with simplicity." Simplicity is elegant.

4:2 Always think in terms of minimum parts and minimum materials.

This is a powerful statement and applies to articles, methods, and apparatuses. Here is an example of the opportunity offered through reducing parts and materials.

A certain electronic component had been produced many years using paper dielectric. It used an impregnant to increase the dielectric constant. A ceramic tube was used to mechanically protect the paper. A wax was employed to seal the tube ends to exclude moisture. Two grommets were needed to anchor terminals to remove stress from the weak paper. In 1920 this was a futuristic idea. The inventor started a new industry and deserves much recognition for the novelty.

By 1950 new plastic films developed for other markets were evaluated to replace the paper dielectric. The plastic film possessed properties which allowed one item to replace six items in the paper construction. The cost reduction was significant while producing a smaller article of higher quality.

Life and reliability favor the reduction of parts and materials. Differential expansion makes a single material preferable unless the differential has an intended function. Single materials avoid dissimilar metal couples, a source of corrosion and failure.

The logistics of procurement and inventory costs clearly give the economic advantage to minimum parts and materials.

The conceptual discipline imposed through minimization brings simplification and longevity to the concept.

4:3 Efficient inventing generally starts with a need. Conversely, finding a need to be filled by a unique technology generally leads to disappointment.

Necessity (need) is truly the mother of invention.

Now and then a new article, method or apparatus surfaces without a need. At first the opportunities for this new solution appear multitudinousness. Gradually, they are ruled out, one by one at great cost and the solution, looking for a need, ends on the scrap pile.

Certainly there are exceptions to this scenario but one must be very careful with these loose solutions. I have never been associated with one which succeeded.

For example:

A nitrogen filled cavity was hydraulically compressed then triggered to drive a hardened steel pin through a 5/8 inch thick steel plate. Why not use this to nail metal together just as soft nails bind wood together? Think of the savings in drilling and tapping to say nothing of the savings of assembly time. Or, use this method to anchor structures to concrete, it has adequate power. Or, fasten steel sheeting to girders in building construction. Or, build truck bodies. All of these requirements use a proprietary fastener, the razor blade, the basis of a business. The driver may even be used in a slaughter house to replace the hammer-blow to the forehead of the victim.

In all applications it offered such coveted features as improved speed, safety, convenience and economy. However, it failed to satisfy all requirements provided by existing solutions and was rejected in each case.

At best, starting with a solution and ending with the need is expensive, time consuming and a perilous course for the inventor.

4:4 Remember, innovation feeds on innovation.

The techniques of inventing become progressively simpler to those who are recognized as inventors. This is not unexpected. It holds for the song writer, the house builder, the physician and the bull-dozer operator. Practice makes perfect. It's made to look easy.

A resident inventor builds a reputation, in the corporate atmosphere. The inventor becomes recognized with an accepted territory. Needs are brought to the inventor by many sources. Help comes from the above and below. Resources are made available.

Selling an invention is generally much more expensive than conceiving and developing the idea. With a track record for inventing and selling other inventions, the inventor receives greater attention than the neophyte selling a concept of equal value. Frequently, an inventor will ride the traveling wave of a new technology and add many inventions along the way. This is especially true if the person invents the basic technology and understands its direction more than anyone else in the world. Success brings success.

CHAPTER 5—PATENTING PATENTABLE IDEAS

5:1 The inventors best friend – the patent attorney.

Patent attorneys usually have a degree in a technical field plus a law degree. They bond easily to inventors. A non-competitive interdependence grows between the two disciplines.

A skilled patent attorney draws many claims from a single patent disclosure. Often the attorney so exhaustively explores and analyzes the claim subjects that even the inventor scarcely understands the legal boundries of the invention.

Without the patent attorney there would be no patents, no patent system and inventions would be protected only as trade secrets. Key ideas would immediately fall into public domain. Incentive for invention would be lost. Commerce, involving new ideas, would become chaotic.

5:2 Write patent disclosures which require minimum effort to interpret into patent applications.

Two patent applications are placed on the patent attorney's desk. The first is neatly typed; each page is numbered, witnessed, dated and signed. The figures are scaled and labeled to teach the idea at a glance. The write-up clearly records the history, other patents, and background leading up to the invention. A description of the invention is detailed with numbered references to the figures. A list of features covers every item which distinguishes this from other teachings and makes this preferred.

The second disclosure is an unintelligible sketch with the message "call me".

These are disclosure extremes and the latter is more common. The patent attorney is concerned, in part, with the number of patent applications he prepares per unit time. Why take his time to chase an inventor when he can be productively employed preparing a patent application.

It's good manners and good business to prepare neat comprehensive patent disclosures.

5:3 Give first priority to protecting novel articles of commerce.

In many cases, *methods* and *apparatuses* are best retained as trade secrets. If the ideas are publically divulged through patents they may teach those who do not respect the patent system. This can involve expensive policing and litigation. However, in cases where methods and apparatuses become items of public use, sale, or manufacture, it is appropriate to consider patent protection.

Some methods and machines are somewhat self-protected by the magnitude of their cost. In other cases, where only a few companies have interest in the technique, trade secrecy is a common practice.

Generally, the *article* is the item to protect if it is publically used in high quantities in a competitive market. The inventor must always consider means for tagging the article, making it simple to spot an infringer. Often the patent numbers are imprinted on the article. In other cases, for various reasons, it is impractical to add the patent numbers.

Frequently an identification naturally associates with the unique property which made the item patentable. Occasionally, the novelty may come from something as subtle as heat-treating. The tag is not obvious except through indirect testing for properties.

During technically sophisticated research, as new stones are being turned, new articles, methods, and apparatuses may be evolving simultaneously. It is particularly important that the innovations be categorized in one of the three classifications to clarify specific areas for protection and avoid the loss of protectable elements.

5:4 Joint inventorship may weaken the legal position of a patent and should be avoided whenever possible.

Most everyone is anxious to see his or her name on a patent. The true inventor may wish to add the name of his pal who works by his side. The boss is sometimes chosen as a co-inventor for many reasons. To keep peace among team members, the team roster is the "et al" which sometimes appears on the patent.

More often someone feels strongly that his suggestion led directly to the invention and insists that he be included as the co-inventor.

The ethics of improperly adding the name of a non-inventor to a patent disclosure should be judged without compromise. In court, a cross examination can often find discrepancies within the testimony of multiple inventors. The legal protection of a perfectly good patent can be lost through this cross examination pattern.

There is an easy solution to the problem of inventorship. If one senses that a subject may involve invention, promptly avoid further discussion. Then write the idea, in detail with sketches as necessary. Sign and date each page. Ask any members who are closely associated with the idea to read, witness and date the write-up. File it away. Repeat if other unique ideas come to mind. As associates contribute a good idea, suggest that they too write it so you can properly witness their contribution.

The above documents should accompany the patent disclosure. As the attorney draws the claims, the attorney can correctly select the inventor or inventors. This procedure avoids misunderstanding among the inventors, but in particular, it provides a much firmer base from which to defend the patent.

CHAPTER 6—CORPORATE INVENTING

6:1 Members of top management who ask for innovative solutions to specific problems can effectively fan the fires of creativity.

Personal recognition causes inventors to seek attention from top management. Mid-management levels should forgive this break of organizational protocol. If the inventor feels identified by the top person, he or she will often apply internal pressure to solve an "insolvable problem". This can lead to insomnia, ulcers, hypertension and other stress related diseases but the aim is contribution, pride and recognition.

For this to work effectively, top management must be committed to innovation. This commitment must carry through every level of the corporate structure. Innovation must be supported as an expression of faith in the company's future, a courageous risk sometimes, but always a planned legacy to those who follow.

6:2 Deep pockets are needed to support uniqueness because it takes time for new ideas to become accepted.

A new product may spend several years in incubation. The inventive flash that gives birth to the idea is generally inexpensive. Preparation of the idea for market sampling may be expensive. The production system and market introduction may be very expensive before the cash flow is positive.

The enormous supporting cost makes the corporation an ideal and in some cases a necessary setting for invention. We hear stories of the entrepreneur who risks everything to cultivate his own idea and build a business. The financial rewards can be higher than those derived from working under the corporate umbrella. But, the personal risk is greater. Available funds are generally less, and there is less time for the inventor to concentrate in his specialized field.

Innovation becomes increasingly expensive as world technology increases in complexity. This drives innovation in to the corporation shelter. The forward looking corporation is sensitive to this direction. The entrepreneur finds a new opportunity developing for his special talent within the corporation.

6:3 Associates, above and below, can make or break an inventor.

It's true; some inventors are odd-balls. Many are individualistic. Others are non-conformists. These characteristics may not have been obvious at the time of hiring. After a few inventive successes an eccentric behavior can develop. This produces an aura which intimidates supervision, causing the inventor to be physically isolated from the daily shocks and routine of business. Finally, the inventor has achieved his ambition. Now he has time to think.

Inventors are desperately needed in the decades ahead. The evolution described in the previous paragraph allows a few inventors to emerge in spite of the fact that they work for an orderly corporate structure. There is another way.

Tomorrow's inventor should resemble a private entrepreneur but using corporate resources. The steps should start with a concept which moves through confirmation, design, fabrication, market sampling, production, selling, improving, diversifying, expanding, then starting over again. With the high technological rate of change, formal training may be required before approaching the next major development.

Supervisors of inventors have the more difficult job. They must listen carefully, lead the inventor without domination, earn the inventor's respect, and never directly contribute to areas which may result in patent claims. In patent sensitive conversations the supervisor should shift to questions rather than statements, leaving the inventor an opportunity to encompass the full novelty. In addition, the supervisor must be capable of leadership in all steps as the concept moves through its metamorphic stages and finally lands in the customer's hands.

Side kicks are of paramount importance to the inventor. Such associates will generally be more highly educated and specialized than the inventor. They enjoy supporting in areas of their specialty. They are far more secure, in personality, than the inventor. Without them the inventor will fail. Their unique contribution should receive special recognition.

The inventor should not be rewarded for extensive administration in preference to invention. If so, he may be pulled into the black hole of management and never invent again.

6:4 Inventors should maintain an active relationship with their company after they retire. Many are reaching their prime at age sixty.

Corporate inventors do not stop thinking at the time of retirement. They search for a worthy cause and drift away. Soon their life style is indistinguishable from one who drove the company truck or guarded the front gate. Nationally, this represents a significant talent loss.

A friend, in his early seventies, retired about ten years ago. He was recognized as an inventor with many significant patents to his credit. He had no specific work plans for the future. He was comfortable.'

Then he became interested in the design of a machine. This led to a second career. Working from a drafting table in his bedroom he engineered a complex assembly line consisting of seven major machines. In addition to mechanical engineering, he became highly involved in lasers, optics, thermodynamics and electronics. His experience guided him through one success after another. A room full of younger engineers could not have approached the success rate and efficiency of this solitary inventor. His efforts have contributed a new industry, receiving international recognition. This project is probably the greatest technical contribution of his career, all happening after he retired.

Printed in the United States
By Bookmasters